YOUR KNOWLEDGE HAS VALUE

- We will publish your bachelor's and master's thesis, essays and papers

- Your own eBook and book - sold worldwide in all relevant shops

- Earn money with each sale

Upload your text at www.GRIN.com and publish for free

Abdul Haseeb

Socio economic conditions: Floral and faunal diversity in the Lulusar Dutipatsar National Park in Upper Kaghan

GRIN Verlag

Socio Economic Conditions, Floral and Faunal Diversity Of Lulusar Dutipatsar National Park Upper Kaghan

Introduction

The word "lulu" means long and "sar" means lake so the lulusar means long lake and it is just because lulusar is the longest lake in kaghan valley. The Lulusar Dutipatsar National park is situated in naran valley 350 km away from mansehra , Khyber Pukhtunkhwa Province, Pakistan. At an altitude 11,200 ft (3,410 m) and located in between Ghatti Das and Baisel villages.

Naran Valley is situated in Mansehra District of Khyber Pukhtunkhwa province. The elevation of the area is 8,125 feet above sea level. It is linked with Mansehra by famous Kaghan route. On the other side it linked with Karakuram highway by Babusar and Botogah routes. The famous Kunhar River originates from the area. The highest peak of the area is Malka Parbat (5290 m) stands in front of the Saiful- Malook Lake. The prominent lakes of the Naran valley are Saif-ul-Maluk, Lulusar and Dudipat.

In April 2003, the Government of Khyber Pukhtunkhwa declared an area of 75058 acres surrounding the Lulusar and Dutipatsar lakes as National Park. They have started management interventions in the area. It is the first national park which is established in private land. Lulusar Dutipatsar is famous for mountain scenery and beautiful lakes lush green pasture and meadows. Lulusar Dutipatsar national park is situated in district Mansehra. The area around Lulusar Dutipatsar is under huge pressure of over utilization and exploitation of natural resources. Forexample Erosion and pollution had tremendously decreased the serenity and beauty of this area, in addition ill managed tourism activities and overgrazing of livestock has increased day by day. Beside, these Alpine Pastures are habitat of many internationally threatened Wildlife species. To arrest habitat degradation and losing beauty of this natural national asset, kpk Wildlife department declared this lake and its watershed as a National Park. This National Park was declared under section 16 of "kpk Protection, Preservation, Conservation and Management Act 1975".on 28 / 03 / 2003 through the (notification number SO (Technical)/VIII-Gen/2003 Government of NWFP).

The Lulusar dutipatsar national park and its surrounding mountains are the the source of rich diversity of flora and fauna having much more importance Snow leopard, Brown bear, Himalayan ibex, Marmot are found within the ecosystem of Lulusar Dodipatsar National park as fauna while Snow partridge (*larwa larwa*), Himalayan Snow cock, eagles are the birds found within the ecosystem of this heaven like area.

There are mainly two types of fish found in the lake (Brown trout, and Rainbow trout). Due to over fishing and various other threats, the rainbow trout is on the verge of extension. People disposed solid waste in the lake, due to which the population of Rainbow trout is decreased similarly people use dynamite in water and electric current causing decrease in trout fish population.

Objectives

1. To enlist the flora of the park
2. To investigate fauna of the park
3. To suggest some Recommendations for Biodiversity Conservation and Eco Tourism.

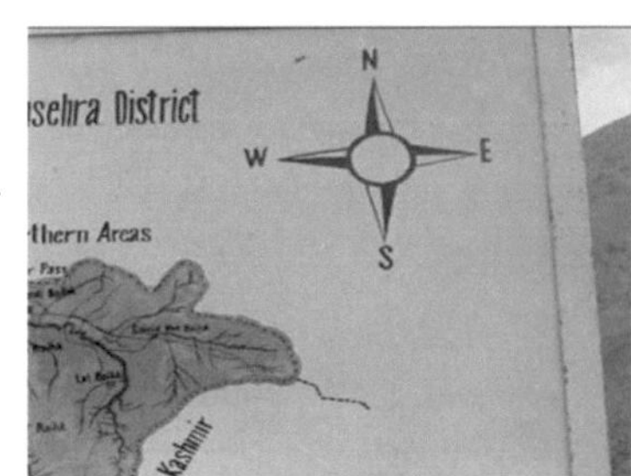

MATERIAL AND METHODOLOGY

Methodology

A survey was conducted in the month of July at Lulusar Dutipatsar national park. The primary data have been collected during the period from 6th July to 12th July in 2011. Primary data was obtained from the local people in order to find out the socio economic conditions,(i-e house hold composition , major source of income ,no of livestock per house etc.) few questionnaires were filled from the local wildlife department staff in the park in order to find the flora and faunal diversity of the park. Questionnaire data was filled from 35 respondents. Some people don't know about the flora and fauna of the park. Plants were collected from different locations (i-e near lake, different villages around the park and from the surrounding mountains of national Park) and were preserved in the herbarium of Hazara University. Plants were identified with the help of the flora of Pakistan. All collected data have been firstly classified and tabulated systematically and then these have been analysed and interpreted. Secondary data have been collected from Khyber Pukhtunkhwa wildlife department official project documents, university library and internet resource.

Material used during survey:

- Binocular was used during survey.
- Map of the park was used for guidance.
- GPS was used during survey.
- Questionnaire was used during survey.

RESULT AND DISCUSSION

Respondent's / Age:

The large proportions of respondents were over 31 years of age. About 40 % of respondents were in the age range of 31-40 years, 31.42 % respondents were from 18-30 years, 8.57 % were from 41-50 year and 20 % were those respondents whose age was more than 50 years.

Martial status:

Analysis about the Martial Status of the respondents shows that high proportion of the respondents 80 % were married while 20 % were unmarried.

Educational Status and level:

The results from the educational status of the overall respondents' shows that 20% respondents were literate and 80% respondents were illiterate

Educational and Level:

The result from the education level of the overall respondent's shows that majority of the respondents about 57.14 % were matriculate while 42.85 % were intermediate .

Occupation and Monthly Income:

Results shows that the highest proportions of the respondents were livestock rearing (45.71%) while 34.28 % were daily wages, 20 % were guid .About 34.28% of the respondents were less than 10,000 income per month. 20% respondents earn income from 10,000- 20,000 per month 45.71 % respondents earn income from 20,000-30,000 per month.

House hold composition

Results shows the highest proportions of children (40.44%) while 30.75 % were males above 12 and 28.8% were females above 12 years.

FLORAL DIVERSITY

5.2.1) Medicinal plant

The Lulusar National Park is rich in biological diversity diversity where wide variety of flora including herbs ,shrubs and medicinal plants are present in all over the national park. The important medicinal plants present in the national park are given below:

Table 5.2.1 Medicinal plants of national park

Family	Scientific Name	Common Name	Local Name
Plantaginaceae	*Plantago ovata*	Greater Plantain	Aspagol
Asteraceae	*Tussilago farfara*	Coltsfoot,	Thandi booti
Berberidaceae	*Podophyllun hexandrum*	Himalayan ayapple	Ban kakri
Primulaceae	*Primula macrophylla*	Primrose	Chikee
Alliaceae	*Allium victrialis*	Jungli piaz	Beshmolo
Boraginaceae	*Onsoma hispidum*	Rattanjot	Kaber
Ephedraceae	*Ephedra intermedia*	Ephedra	Shalako
Fabaceae	*Trifolium repens*	White clover	Telsharong
Aseteraceae	*Taraxacum officinale)*	Dandelion	Paptra
Asteraceae	*Artemisia maritime*	Artemisia	Makhoti

Shrubs

These are found in alpine pastures/tree line limits therefore no woody vegetation or no dense woody vegetation is found in the park there are varieties of shrubs found in the park such as ,

Table Shrub of the National Park

Family	Scientific Name	Common Name
Grassolariaceae	*Ribes alpestere*	Alpine gooseberry.
Cupraceae	*Juniper exelsa*	Grecian Juniper
Tamaricaceae	*Tamarix gallica*	French marisk
Betula utilis	Betulaceae	Jongi
Barberidaceae	*Barberis achycantha*	Sumbal
Crassulaceae	*Rasularia alpestris*	Not know
Rosaceae	*Rosa wibbiana*	Paul wari
Cupraceae	*Juniper communis*	Juniper

Table Flora of the Lulusar Dutipatsar National Park

Botanical Name	Local Name	Family	Type of vegetation
Polygonum paronchioides	Chutyal	*polygonacea*	Herb
Bistotrta affinis	Jogi badshah	*polygonacea*	Herb
Oxyria digyana	Tindi jari	*Polygonacea*	Herb
Thymus linearis	Banjameri	*Labiatae*	Herb
Nepta discolor	Patmea	*Labiatae*	Herb
Phlomis bracteosa	Ghorie	*Labiatae*	Herb
Mysotis alpestris	Peli panja	*Borigenaceae*	Herb
Potentilla dryandanthoides	Chalandri	*Rosaceae*	Herb
Pseudomertentia echioides	Not know	*Borigenaceae*	Herb
Poa alpine	Ghas	*Poaceae*	Herb
Trifolium repenses	Saag	*Mimosaceae*	Herb
Polygonum coagnatum	Masloon	*polygonacea*	Herb

Tuxilage farfora	Thara bori	*Astraceae*	Herb
Shibbaldia purpacra	Not known	*Rosaecae*	Herb
Salvia longifloia	Kaljari	*Lamiaecae*	Herb
Chorispora bungena	Not known	*Crucifere*	Herb
Dactylorhiza hatagerieae	Not known	*Orchidaceae*	Herb
Iris hokiriana	Gora grass	*Iridaceae*	Herb
Phlomi rotate	Not known	*Borigeneaceae*	Herb
Viloa biflora	Banafsha	*Violaceae*	Herb
Lotus corniculata	Pili boti	*Pepleonaceae*	Herb
Pedophylum hexandrium	Ban kakri	*Podophylaceae*	Herb
Caitha alba	Not known	*Renuculuaceae*	Herb
Primula macrophylla	Klimire	*Pimulaceae*	Herb
Chesyena nubigeria	Not known	*Peplionaceae*	Herb
Chesnya nusigeria	Not known	*Pepionaceae*	Herb
Trixicum efficineles	Hund	*Asteraceae*	Herb
Rumex aestosa	Hola	*Polygonaceae*	Herb
Aritmeasa santolinifolia	Chow	*Astraceae*	Herb
Selinum tenecifolium	Not known	*Selaginellaceae*	Herb
Dryopterius sinofibrillosa	Kunji	*Pteridaceae*	Herb
Fragaria nubicola	Not known	*Rosaceae*	Herb
Achitia melifolium	Not known	*Accraceae*	Herb
Adiantum reustum	Ball	*Tapteridaceae*	Herb
Onoprdum acanthium	Not known	*Astraceae*	Herb
Anagallis arrensis	Not known	*Primulaceae*	Herb

FUANAL DIVERSITY

Lulusar Dutipatsar national park is rich in biological diversity where wide variety of wild animals, birds and migratory birds are found.In Lulusar Dutipatsar national park there are two types of fishes found in the lake . Some of the important fauna of the park is given below .

Wild Mammales

Important wild mammles of the Lulusar Dutipatsar national park are Himalayan ibex(*Capra ibex* sibirica), Brown bear(*Ursus arctos),* Marmot*(Marmot flaviventris)* snow leopard**,** (*Uncia uncia)*

Table Wild Mammales of Lulusar Dutipatsar National Park

Scientific Name	Common Name	Family	Status
Capra ibex sibirica	Himalayan ibex	*Bovidae*	Native
Ursus arctos	Brown bear	*Ursidae*	Native
Marmot flaviventris	Marmot	*Sciuridae*	Native
Uncia uncia	Snow leopard	*Felidae*	Native

Birds

Important bird species of Lulusar Dutipatsar national park includes:

Snow cock (*Tetragallus himalayensis),* Snow partridge (*Lerwa lerwa),* Himalayan Griffen vulture (*Gyps coprotheresand),* Himalayan Monal *(Lophophorus impejanus)* and Falcon(*Falco jugger*)

Table Birds of the National Park

Scientific Name	Common Name	Family	Status
Tetragallus himalayensis	Snow cock	*Phasianidae*	Native
Lerwa lerwa	Snow partridge	*Phasianidae*	Native
Gyps coprotheresand	Himalayan Griffen vulture	*Accipitridae*	Native
Lophophorus impejanus	Himalayan Monal	*Phasianidae*	Native
Falco jugger	Falcon	*Falconidae*	Native

Migratory Birds

Some the of the migratory birds come to Lulusar Dutipatsar national during autumn and spring season. Important migratory bird species are Pintail (*Anas Acuta*), common teal (*Anas crecca*), mallard *(Ans platyrhynchos).*

Table migratory birds of national park

Scientific name	Common name	Family	Status
Anas Acuta	Pintail	*Anatidae*	Migratory
Anas crecca	Common teal	*Anatinae*	Migratory
Ans platyrhynchos	Mallard	*Anatidae*	Migratory

Fish

There are mainly two types of fishes found in the Lulusar Dutipatsar national. Brown trout (*Salmo gradnarie*) rainbow trout(*Salmo truta fario*) .These fishes survive in very cold water .

Discussion

Based on results the socio economic condtions of the local people living in the park area shows that most of the people in national park area were migratory herds , they came to the park during summer season i.e (June –August) where they graze their livestock on pastures .The main source of income is livestock rearing and tourism , average family of the household were 10 persons per household , 80% of the people were illiterate 90% of the people were living in kucha houses , there were no basic facilities for local people .Lulusar dutipatsar national park is famous for natural beauty and fresh water springs and lakes. Total 54 plant species were collected in which 35 species were herbs ,10 medicinal plants and 9 species were shrubs. Due to increase pressure of human and livestock the flora of the park were decreased .Based on the results it was concluded that government should take immediate action on priority basis to conserve the biodiversity of national park .

REFRENCES

(wwfpak report, 2006) Field survey report) Prominent Naran Lakes (Saif-ul-Maluk, Lulusar and Dodipat)

10. Rabnawaz (2009) economic valuation of tourism based on lulusar lake in Kaghan valley of district Mansehra thesis .